A CRYPTOZOOLOGICAL STUDY
OF THE
SHUNKA WARAK'IN

ISBN: 1-4392-1657-6

EAN13: 9781439216576

To order additional copies, please contact us.

Booksurge Publishing
www.booksurge.com
1-866-308-6235
orders@booksurge.com

Table of Contents

- BOOK 1 -

CRYPTOZOOLOGY:
IS IT SCIENCE OR SCIENCE FICTION?

Introduction

What is cryptozoology?

The word cryptozoology is derived from the Greek word "kruptos", meaning "hidden" and the English word "zoology" which means the scientific study of animals. Thus, cryptozoology literally means the "study of hidden animals."[1]

Cryptozoological animals (a.k.a. cryptids) fall outside of contemporary zoological classifications. These animals consist primarily within two fields of research[4, 6]:

- The search for living examples of animals which are believed to be extinct.
- The search for animals that have been reported as being sighted, but no physical evidence has surfaced to confirm the existence of the creature.

Some of the more famous cryptids include the following: Bigfoot, Sasquatch, The Yetti, The Yowie, The Lock Ness Monster ('Nessie'), and the Chupacabra[6].

Unfortunately, cryptozoology is not generally recognized by the scientific community as a true science. It is commonly believed that since physical specimens of the creature in question have not been regularly seen and have never been examined that it must not truly exist, and those individuals that have reported spotting the creature are merely suffering from hallucinations or seeking attention.

Generally the world of science is associated with that which can be proven to exist because the data can be classified and quantified. On the other hand, cryptozoology is often related to the world of fiction, superstition, and fantasy, and, therefore, does not exist until sufficient proof surfaces to prove otherwise.

This work attempts to view both sides of this debate and discuss weather or not to classify cryptozoologists as scientists.

The Scientific Method

Science places a great deal of importance on what is termed the "scientific method." In fact, only fields of science that adhere to this method are actually considered legitimate by the scientific community. So, since this process is so significant, what follows is an explanation of the scientific method in detail and the strengths and weaknesses of this concept.

The scientific method describes a systematic process that is used to ask and answer scientific questions by making observations and conducting experiments [2]. There are six stages to the scientific process:

1. Ask a question
2. Do background research
3. Form a hypothesis
4. Perform an experiment to test the hypothesis
5. Analyze data and make a conclusion
6. Communicate your results

Applying the scientific method alone to solve a problem or to make new discoveries does not constitute a scientist. After all, if this were true, we would all be scientists. Here is an example of applying the scientific method to solve a common problem:

1. Ask a question – Why does the television not turn on?
2. Do background research – Consider various reasons why the television may not turn on.
3. Form a hypothesis – The television does not turn on because it is not receiving electrical power.
4. Perform an experiment to test the hypothesis - Plug the power cord into the wall outlet and attempt to turn on the television by depressing the power button.
5. Analyze data and make a conclusion – Since the television was disconnected from a wall outlet, the television did not receive electrical power and failed to operate.

4

6.   Communicate your results – "Hey kids, I got the television to work by plugging it into a wall outlet."
Congratulations, you are a scientist!

In this example, the first hypothesis of why the television failed to turn on was correct and the problem was immediately resolved. Since this solution is theoretically valid and continues to solve the problem under each case of similar occurrences, the hypothesis graduates to a theory [8]. Since this theory holds true for each instance that the television is unplugged from the wall, the theory becomes recognized as a scientific fact.

This condition does not, however, explain the only reason why a television may fail to turn on. There could be any number of reasons why the television may not turn on; these reasons may include (but are not limited to) the following:

1.   A faulty power switch
2.   A blown house fuse
3.   A blown television fuse
4.   A fault at the electric company or generator
5.   Faulty house wiring

Thus should any television fail to operate, even though it is connected to a wall outlet, this does not prove the theory or fact inaccurate, only incomplete. Further hypothesizes and testing would need to be conducted to derive every possible reason of why the television fails to operate correctly.

A common practice used by scientists to test a new drug or nutritional supplement is to administer the real supplement to a large group of individuals (test group) and a placebo (fake) to a different group (control group) and monitor the results of each. The measured change in the results of the members of both groups, however, may not be directly related to the drug itself, but by the physical condition and mental attitude of the individual subjects. For example, suppose that a new sports enhancing supplement is to be tested. The real supplement is given to 100 athletes in the test group and the 100 control subjects took a placebo. There may be individuals in each group that may work out harder and with more intensity because they believe that they are being 'pumped-up' with a new miracle pill. Others, however, may tend to step down their training efforts feeling that the new miracle pill would 'fill- in' for their reduced training intensity.

Then, after a predetermined training period, the strength of the subjects of each group are measured and compared to describe the effectiveness of the new supplement. The results are typically presented as follows:

'Of the 100 participants that were administered the test supplement, 61% showed a 10% increase in strength, 26% showed no change in strength, and 13% showed a 1% decrease in strength. However, of the 100 participants that were administered the placebo, 30% showed a 8% increase in strength, 50% showed no change in strength, and 20% showed a 3% decrease in strength. Thus, since 61% of the athletes showed a 10% increase in strength over the 8 week test period, the new supplement is a proven success.'

Although the scientists attempt to regulate the exercise under consideration (i.e. the bench press) to be constant for every subject, nothing is generally mentioned about each member's physical activity level or previous health levels outside of the laboratory.

Not even science is an exact discipline. Practically every hypothesis, theory, or fact may eventually need to be re-evaluated to account for a newly discovered condition that disobeys previous avenues of thought. Test and control groups in an experiment are subject to unknown variances that could cause the test results to go awry. Thus, with all of the variances and inaccuracies that occur in science in general, why should cryptozoology be criticized as not being a true science?

The Cryptozoologist's job is much more difficult than that of a typical scientist due to the scarcity of encounters and the smaller pool of subjects. Cryptozoologists are often left with insufficient evidence (such as only a footprint) to indicate the presence of the cryptozoological animal and then must hypothesize about the creatures' nature by what can be speculated from this limited source. Additionally, with the cryptozoological creature considered as an endangered species and rarely seen, the probability of gathering 200 of these animals together to conduct a controlled experiment would prove highly unlikely.

Types of observations

Within the world of science, there are 3 major types of observations: tangible, testimonial, and anecdotal [1, 8].

Tangible evidence is a physical specimen or the remnants of one (such as a bone, tooth, or hair samples), which provides undeniable evidence of something's existence. The primary goal of the cryptozoologist is to obtain tangible evidence from a cryptid to prove its existence.

Testimonial evidence relates to second-hand information, such as photographs or reports. This type of evidence is not conclusive because it cannot be directly analyzed. For example, you may see a picture of a Bigfoot, but by viewing the picture alone you would be unable to conclude if the image is authentic or merely a person in an ape suit [8].

Anecdotal evidence is, basically, "eyewitness accounts." This evidence is infamously unreliable. However, basic characteristics as to the creatures' physical appearance and demeanor can generally be accepted as true if similarities can be discovered among several eyewitnesses (particularly if they are not being coached into describing the creature a certain way in an attempt to meet the expectations of the investigator).

Why Cryptozoologists should be viewed as scientists

A good cryptozoologist conducts their investigations using applicable scientific methods and psychology to determine if a valid case exists and if it is worth investigating. Cryptozoologists are fully aware that the possibility of hoaxes exists and should take care that they are not led onto a 'wild goose chase' for the amusement of the hoaxer(s). For example, should a cryptozoologist find a footprint of a cryptid (such as from a Bigfoot creature), careful scrutiny would normally first be employed to ensure that it is authentic. Inspection of the footprint could include (but is not limited to) the following observations:

1. The depression of the footprint – The depth of the footprint indicates the approximate weight of the creature. If the approximated weight is roughly equivalent to that of an average human, the print may be a fake.

2. The condition of the surrounding soil – Does it look chipped away; is there a strange residue surrounding the perimeter of the footprint? These could all be indications of a man-made hoax.

3. Strange geometric shapes within the footprint – If the footprint contains a perfect circle or an equilateral triangle it could be man-made. On the other hand, the creature may have stepped upon some perfectly shaped object that created a perfectly symmetrical scar upon his foot.

4. Unsymmetrical physiology – Generally all known creatures possess symmetrical body proportions (i.e. our left little toe is just as big as our right little toe). If the footprints possess unsymmetrical properties, there could be a high probability that the footprints were man-made using non-exact fabrication processes.

Regardless of whether the footprint is a suspected fake or not, a cast of it should still be created for later study. This procedure fulfills the collecting and analyzing data components of the scientific method [3]. By analyzing the characteristics of the casted footprint, a determination as to the authenticity of the actual footprint can be made, and if the sample is suspected to be genuine, a hypothesis can be formulated as to the proportions and weight of the whole creature.

Unfortunately, without a physical specimen for analysis, the testing of the hypothesis and the reporting of the results would not prove practical, thereby, failing to fulfill all stages of the scientific method. Thus, it can be argued that cryptozoology cannot be classified as a true science. However, the scientific method is not all-inclusive, and it also is subject to inaccuracy. For example, if I stepped in the sand and you made a cast of my footprint, you could speculate as to my physical weight and proportions. But, if you were never to see me, your hypothesis as to my characteristics would remain unsubstantiated. Furthermore, just because you may never see me neither means I do not exist, nor can evidence of my existence be denied!

Interview skills

To obtain the truth regarding a cryptid encounter, a cryptozoologist must be a skilled interrogator. This does not mean one should play detective and shine bright lights into the interviewees' eyes or play "good cop/bad cop" to obtain the truth about the encounter. However, the researcher should be careful not to lead the interviewee into an inaccurate recollection of their encounter to coincide with the expectations of the investigator. Furthermore, the researcher should be aware of the interviewees' body language and eye movements. If the interviewee rarely makes eye contact during the interview, he may be lying about the encounter.

A useful practice to include in the cryptozoologists' toolkit during an interview is the eye accessing technique [10]. This technique relies upon the principal that different parts of the brain are activated depending upon the type of information being accessed. By watching the movement of the eyes, the interviewer can detect if the interviewee is accessing visual memories (that which is being recalled from memory) or creative ability (that which is being created using the imagination) to answer a question. The location in the brain that stores each type of information is dependant upon each individual's dominant hand. For a right-handed person, the eyes go up and to the left when accessing visual memories (and up and to the right for a left handed person), and go up and to the right when accessing the imagination (up and to the left for a left handed person) [10].

During the interview, be conscious of any double talk that may occur to determine if the interviewee is relating the true encounter to you as he experienced it or making it up as he or she speaks. By recording the conversation, you can review the interview several times in an attempt to find any incongruities in the descriptions.

During an interview with an individual whom claims to have spotted a cryptid, the cryptozoologist should ask the interviewee several open-ended questions about the encounter [1]. These questions need answering using as highly detailed descriptions as possible. Questions that require a simple "Yes" or "No" response should be avoided to prevent acquiring inaccurate data and leading the investigation astray. Consider this situation:

Suppose that the interviewee actually observed a large ape and has misidentified it as being a Bigfoot; should the cryptozoologist be overly zealous in tracking down a real Bigfoot, he may ask Bigfoot characteristic questions only. The interviewee would then be eager to agree with the cryptozoologist and confirm his questions with a positive "Yes," thereby causing inaccurate data to be collected which would lead to an inaccurate hypothesis to be formulated.

Although people, interview, and body language skills are considered more of an art than as a science, they are still integral parts of the cryptozoologists' tools; these skills should not be dismissed as being useless or impractical. Furthermore, despite popular beliefs, a cryptozoologist must be well educated in wilderness survival, animal behavior, hunting & tracking, scientific principles, evolutionary theory, recordkeeping, and other abilities which require a greater than average mental ability to perform.

Why Cryptozoologists should not be viewed as scientists

One of the prevailing arguments supporting why cryptozoology should not be viewed as a true science is because there is no standardized curriculum of study for this field [3]; since this field of study is comprised totally of the unknown, no one is qualified to write a worthy textbook on the subject matter [5]. Generally, however, all non-fiction books are based upon research. It is doubtful that any author simply sits down and writes a non-fiction title about a cryptid without first conducting the proper research into the subject matter. After sufficient data has been researched, ideas and beliefs that are unknown about the cryptid can be logically reasoned from analyzing the collected data and applying associations from their suspected related species.

It has also been argued that since degrees in cryptozoology are not generally offered, that this field of study neither exists nor has merit [3]. This idea rests in the belief that one must be recognized or accredited and must also work within a profession to be accepted by the profession.

Generally science is considered valid if we can perform tests upon a tangible specimen and invalid if no tangible specimen exists. This concept holds great value upon earthbound specimens, since these entities can be readily gathered and studied, but what about the field of theoretical physics? Although experimentation cannot be performed within this field of study, theoretical physicists still try to make advances within this field through the use of believable mathematical models [9]. Astrophysics books covering topics such as black holes, warm-holes, parallel dimensions, and other spatial anomalies are regularly published, and yet no one appears to deny this particular branch of science. Since one cannot place a black hole under a microscope for analysis and experimentation, it must be assumed that this field of study is highly theoretical and all of the conclusions on this subject matter are only speculations. Similarly, cryptozoology is based upon forming hypothesizes from the little evidence collected from a particular cryptid and basing assumptions upon what would logically be considered as an associated species. One must then question why one branch of theoretical study is highly respected and accepted and another one is ridiculed as science fiction.

It seems rather contradictory that science claims to be willing to explore new avenues but is unwilling to do so until proof that the new avenue actually exists comes to surface. Natural radio frequency transmissions, electricity, and atomic activity have existed since before the dawn of man but have only been recently accepted, studied, and put to productive use. In this regard, scientists seem to have a type of tunnel vision and can only accept what they can see directly in front of them; that which exists within their peripheral vision does not exist until it comes into their primary focal point. Thus should a cryptid ever be brought before the scientific community for analysis, the new creature would only then be eagerly accepted and studied. In time, new textbooks would be published, university classes would be made available, and a whole new science dedicated to the research of this new creature would be created. But even though the discovery and study of this cryptid should establish the field of cryptozoology as a valid science, scientists would claim the newly discovered cryptid as merely a known animal with a unique genetic defect and still claim cryptozoology as a type of science fiction unless new cryptids could be *routinely* discovered and analyzed.

If this avenue of thought is accepted, only when the researcher is successful in making a significant scientific discovery that progresses science or aids mankind is he or she considered a true scientist. By this account, all of the data collection, analysis, theories, and experiments conducted within the field of science merely labels an individual as a scientific hobbyist or 'tinkerer' until the scientific breakthrough is finally achieved.

Generally, cryptozoologists are seen merely as hunters who are on a never-ending expedition chasing after the ever-elusive cryptid [3]. It is believed that the cryptid does not exist and the hunter is merely wasting his time. If the cryptozoologist goes to the field merely with the 'See it & Shoot it to prove that it exists' mentality, then he is merely just a hunter and neither a true cyptozoologist nor a scientist in any respect. However, should the cryptozoologist use his education and training to make believable speculations and formulate hypotheses about the cryptid, then he should consider himself both as a cryptozoologist and a scientist [7].

Summary

The word cryptozoology literally means the "study of hidden animals." This field of study pertains to the search for animals (a.k.a. cryptids in the field of cryptozoology) that primarily fall within two fields of research:
- Animals that are believed to have survived extinction.
- Animals that have been reported as being sighted, but no physical evidence has surfaced to confirm their existence.

As no tangible evidence of a cryptid has ever been scientifically examined, the existence of the animal lies in debate. Thus, cryptozoology is not generally recognized by the scientific community as being a true science.

The scientific world places a great deal of importance on what is termed the scientific method. There are six basic stages to the scientific method:

1. Ask a question
2. Do background research
3. Form a hypothesis
4. Perform an experiment to test the hypothesis
5. Analyze data and make a conclusion
6. Communicate your results

Unfortunately, only fields of science that adhere to the scientific method are actually considered legitimate by the scientific community. This rule of conformity makes it difficult for the cryptozoologist to claim his profession as a true scientific field of study. Whereas science generally has a plethora of specimens they can examine and conduct experiments upon, the cryptozoologist generally possesses little in the way of physical evidence and knowledge of the cryptid under investigation and must use his education and experience to make speculations and postulate hypothesizes as to the nature and physical characteristics of the hidden animal. Regrettably, the cryptozoologist is often unable to test the hypothesis because there may be no physical specimen for examination and experimentation.   This fact alone renders cryptozoology questionable as a true branch of science.

The three major types of observations recognized by the scientific community are tangible, testimonial, and anecdotal. Tangible evidence is a physical specimen or remnants of one (such as a bone, tooth, or hair), which provides undeniable evidence of something's existence. Testimonial evidence relates to second-hand information, such as photographs or reports. This evidence is not conclusive because it cannot be analyzed directly. Anecdotal evidence is, basically, "eyewitness accounts" and is infamously unreliable.

Good cryptozoologists conduct their investigations using applicable scientific methods and psychology to determine if a valid case exists and if it is worth investigating. Knowing how to identify cryptozoological hoaxes through reason and scientific investigation helps to eliminate unnecessary attention paid to false evidence. Consequently, the cryptozoologist must be well educated in a wide array of different talents, some of which would include the following:

1.   Hunting, tracking, and trapping
2.   Wilderness survival
3.   Hunting & firearm regulations
4.   Specimen and data collection
5.   Record keeping
6.   Animal tranquilizers
7.   Basic first aid
8.   Animal behavior
9.   People, interview, and body language skills

It can be reasoned that if a cryptozoologist applies scientific principles to his or her studies and findings, then he or she is performing the role of a scientist and should then be considered a scientist.

Since the field of cryptozoology is highly associated with the unknown, it is speculated that few possess adequate knowledge on the subject and are qualified to write about it. This opinion, however, appears unsubstantiated, as generally the authors have read all of the reports, collected sufficient historical data, and based their speculations on associated animal behavior and characteristics prior to publishing their work.

Generally science is considered valid if we can perform tests upon a tangible specimen and invalid if no tangible specimen exists. In the field of theoretical physics, mathematical models are constructed to represent an event that is not validated through experimentation. Similarly, a cryptozoological hypothesis is primarily based upon testimonial and anecdotal evidence as well as animal behaviors, which are assumed to be related to the cryptid under investigation. Thus, although neither field of study provides a tangible specimen for experimentation, both fields of study should be viewed as a science provided that acceptable scientific explanations are provided to substantiate their conclusions.

Conclusion

It appears that the biggest restriction to cryptozoology being classified as a true science is no tangible evidence exists to confirm the reality of the cryptid. Without a physical specimen to examine, no experiments can be performed upon the creature, and thus any hypotheses about the creature remain invalidated.

Although the cryptids under direct consideration are classified as unknown, they can generally be associated with similar known species extant or extinct. By applying the characteristics of similar species to the cryptid in question and analyzing historical sighting reports, the cryptozoologist can make a hypothesis about the physical and behavioral characteristics of the unknown cryptid. The cryptozoologist can then alter their hunting pattern or procedure in accordance to the newly created hypothesis. Should the hypothesis prove correct, the cryptozoologist may be one step closer to obtaining an actual physical specimen of the cryptid. By applying these scientifically based principals to the field of cryptozoology, this study should receive the same amount of support and attention as any other branch of theoretical study.

References

1. CRYPTOZOOLOGY SCIENCE & SPECULATION, Author: Chad Arment, Copyright 2004 by Chad Arment, Published by Coachwhip Publications, ISBN: 1-930585-15-2

2. Steps to the Scientific method – http://www.sciencebuddies.org/mentoring/project_scientific_method.shtml

3. What is Cryptozoology? - http://www.cryptozoology.com/articles/wic.php

4. Cryptozoology - http://en.wikipedia.org/wiki/Cryptozoology

5. Is cryptozoology a valid science? - http://www.helium.com/debates/158588-is-cryptozoology-a-valid-science/side_by_side

6. Online Class: Cryptozoology 101 - http://www.universalclass.com/i/crn/7550037.htm

7. Would you consider cryptozoology a real science? - http://au.answers.yahoo.com/question/index?qid=20080426230041AA2Xkwe

8. CRYPTOZOOLOGY - http://silentmoviemonsters.tripod.com/TheLostWorld/LWCRYPTO.html

9. Theoretical physics – http://www.en.wikipedia.org/wiki/Theoretical_physics

10. Instant Fact: How To Get The Truth Out of Anyone!, Never Be Lied To Again Compiled By: John J. Webster, By David J. Lieberman, Ph.D., St. Martin's Press, New York 1998, DD: 158.2, ISBN: 0-312-18634-7

- BOOK 2 –

A CRYPTOZOOLOGICAL STUDY
OF THE
SHUNKA WARAK'IN

Introduction

The Shunka Warak'in (Native American for "carries off dogs") [2,4,5] is a primitive wolf-like beast exhibiting hyena-like characteristics. This creature is believed to inhabit the wilds of the upper Midwestern United States. It is rumored to be a carnivorous predator that viscously attacks livestock animals. Unlike the surgical precision that a modern wolf may use to take meat from a domestic herd, the Shunka Warak'in savagely attacks its' prey (a behavior characterized by a primitive animal) and leaves several animals wounded during its feeding frenzy. Thus far, there have only been three suspected sightings of this cryptid.

First Reported Sighting

The first Shunka Warak'in sighting is purported to have occurred in 1886 by the Hutchins family [8, 9]. Shortly after they settled within the Madison River Valley in the lower part of Montana, the Hutchins family, as well as other locals, had encounters with an unusual wolf-like creature. It was described as nearly black in color and with high shoulders and a downward sloping back, much like a hyena [1, 7]. This strange animal was rumored to be a great traveler, as its appearance had been spotted at the main ranch and at other ranches ten to fifteen miles away [1,7].

One morning in late January the Hutchins grandfather, Israel Ammon Hutchins, was alerted to the beasts' presence by the barking dogs [1, 2]. Israel responded by shooting and successfully killing the creature. The carcass of the Shunka Warak'in was then donated to Joseph Sherwood, who displayed it as a taxidermal specimen for many years in his combination grocery store and museum at Henry Lake in Idaho [1]. Mr. Sherwood named this new exhibit "Ringdocus." [1, 2, 4, 8] This only known piece of evidence that confirmed the creatures' existence went missing when it was moved to an unspecified location in the West Yellowstone area. Finally, in December of 2007 this evidence resurfaced in an exhibit at the Idaho Museum of Natural History in Pocatello. This rediscovered find measures 48 inches (122 cm) in length as measured from the tip of its snout to its' anus, not including the tail, and stands 27-28 inches

(69 – 71 cm) high as measured from the front paw to the shoulder [9]. It has a lupine head and a narrow snout [9]. A sloping back and short hind legs characterize the posterior of the animal. Its' coat consists of a dark brown to black colored fur with some lighter tan colored areas. The coat also displays faint impressions of stripes on its' flanks. In general, the cryptid resembles a cross between a wolf and a hyena. This animal has since been given the scientific name "Guyasticuts" (Rocky Mountain Hyena).

It is believed that DNA testing will reveal the nature and origin of this cryptid, but the results of the tests are yet to be published.

Additional Encounters

In 1995, creatures similar to the Shunka Warak'in were sighted in Nebraska, Iowa, and Illinois in Northern America, and Alberta, Canada [1, 7]. Lance Foster, an Ioway Indian, reported that a strange animal that he called a Shunka Warak'in, which resembled that of a hyena, snuck into his tribes' camps at night to steal dogs. He stated that it cried like a person when they killed it [1, 7]. The remains of this creature are rumored to be in the possession of an unknown person [1].

In December 2005, a large, unusual wolf-like animal started killing livestock in the Montana counties of McCone, Garfield, and Dawson. By March, 2006, this creature had been reported to have attacked six herds of sheep in McCone and Garfield counties. By this time the creature had been accused of wounding 71 and killing 36 ewes [10], earning it the nickname "The creature of McCone County." By the end of October 2006, the creature had 120 livestock killings attributed to its' name.

On November 2, 2006, the animal assumed to be the creature of McCone county was killed from the air in Garfield county by Montana's Wildlife Service agents [7]. The animal that was shot weighed 106 pounds and was covered in shades of orange, red, and yellow fur. These color variations differed from those of the Northern Rockies wolves, whose fur is characterized by color shades of gray, brown, and black.

Muscle tissue extracted from the animal was sent to the University of California in Los Angeles for DNA testing, and the carcass went to the National Fish and Wildlife Forensics Laboratory in Ashland, Oregon for genetic study[7].

Although the Montana Wildlife officials were unable to identify this particular creature, it has initially been declared by an article on the Montana Fish, Wildlife, and Parks website to be a four year old male wolf with uniquely colored fur[2].

Later however, in an article posted on outdoornewsdaily.com dated September 02, 2008, a re-examination of the tissue sample at the UCLA laboratory revealed a mixture of genes which are not found in naturally-occurring wild wolf populations[11]. Laboratory results confirmed that the DNA sample was extracted from a wolf-dog hybrid created from the result of a private captive breeding experiment [11]. Furthermore, the laboratory was able to confirm that this hybrid creature was responsible for the killing of approximately 100 sheep in the Garfield and McCone counties in 2006, thereby revealing the true identity of "The Creature of McCone County."

Montana law requires that any captive, domestic, or hybrid wolf which is more that half wolf to be permanently tattooed and registered with the Fish, Wildlife, and Parks (FWP) foundation [11]. Additionally, should any escape, release, or other change in disposition of these animals occur, state law requires that the FWP be notified [11]. Since this particular hybrid creature was neither tattooed nor the FWP ever notified of its' existence, it may be implied that the person(s) involved with this experiment assumed that the necessary authorization to proceed with (what might be considered) an unethical or illegal genetic experiment would not be acquired and thus attempted to conduct the experiments in secret.

Before accepting "The Creature of McCone County" as dead and the case is closed, there are a few interesting questions to consider:

- Does a link exist between this creature and the suspected Shunka Warak'in sightings of 1995?
- How long did this hybrid creature exist in the wild?
- What impact did this new animal have on the ecosystem?

- With what animals did this hybrid successfully mate with, and what characteristics may the offspring develop?

These are important questions that may have been selectively "overlooked" to keep the populace calm and prevent an inrush of 'hybrid monster' reports, or are still being investigated.

As a result of this incident, a recent federal court decision reinstated the Endangered Species Act protection for wild wolves inside the northern Rocky Mountain region [11]. This ruling proclaims that the wolf population that exists within the northern part of the state to be declared as "endangered," and that which exists within the southern half of the state as "non-essential experimental" subjects.

According to this ruling, the endangered wild wolves of northern Montana cannot be hazed, harassed, or killed by livestock owners or other private citizens [11]. On the other hand, the wolves of southern Montana can be hazed, harassed, or killed if they are witnessed attacking livestock or domestic dogs on either private or public land. The incident, however, must be reported to the FWP within 24 hours.

Although this experimental wolf-dog hybrid explains away the mystery behind "The Creature of McCone County" and may also account for the 1995 suspected Shunka Warak'in sightings, it is highly unlikely to also explain the creature mounted as the 'Ringdocus' exhibit.

Possible 'Ringdocus' explanations:

It may be assumed that the 'Ringdocus' creature was most likely one of six different possibilities:

1. Merely a poorly mounted wolf
2. A hand-stitched creation
3. A unique genetic anomaly
4. A truly successful natural hybrid
5. Other
6. A surviving prehistoric or Ice Age creature

Although it is possible that Joseph Sherwood may have been an inept taxidermist [5, 6, 8], this explanation is highly improbable. If Mr. Sherwood ran a taxidermal museum, it would be logical to assume that his exhibits were all properly mounted for display, for if the exhibit was not properly mounted, his reputation would be tarnished and he would loose the interest of customers visiting his museum/store combination.

It is however, likely that he purposefully altered a standard specimen to create a new animal to draw new customers into his place of business. For example, by removing specific vertebral discs from the spine of a wolf, and then wiring the remaining pieces together, a sloping back can be given to a standard wolf carcass. Then, the rear legs could be slightly retracted within the body to give the animal the appearance of shorter rear legs. Finally, selected dyes could be used to give the "new" creature a different hide appearance. This newly created animal could then be advertised as a new zoological discovery to draw customers into his business to see the newly discovered exhibit and to spend their money. Later, after sufficient attention has been given to this discovery, an eminent professor (such as from the field of paleontology) may have wished to view the exhibit and make conclusions upon its authenticity. Mr. Sherwood, knowing that his exhibit could not withstand the scrutiny of a professional, conveniently had his exhibit go missing. Thus, with no evidence to be examined, the claims about the creature could not be proven; thereby leaving his customers misinformed and satisfied that they had witnessed a truly unique exhibit. This course of action would surely prove more beneficial to Mr. Sherwood than facing an angry mob and being accused of 'swindling' his customers out of their money.

Now advance forward in time about 120 years. This time the 'Ringdocus" exhibit is recovered from some collected junk (such as from an attic, for instance), from someone completely unfamiliar with its history and the 'hype' begins anew. Actually, the act of stitching together parts of various animals to create a "new" creature is an old practice. One commonly faked creature was a mermaid skeleton comprised of the upper region of a monkey and the tail end of a fish[12]. Perhaps the "Ringdocus" was created from stitching the front end of a wolf to the hind section of a hyena to "create" a new creature.

It is possible that the "Ringdocus' was the natural offspring of a typical animal born with a unique genetic defect. For instance, an unbalanced growth hormone may have circulated through a common wolf's premature fetus. This situation may have caused an accumulation of growth hormone toward the front of the body (which may have lead to a more aggressive personality) and a reduction of growth hormone toward the hind section of the creature (leading toward lesser developed hind legs). This assumption is, of course, just a random guess. However, a thorough inspection and medical analysis of the recovered creature may provide evidence to support or deny this belief.

Although it is commonly considered unlikely or impossible by today's scientific knowledge, a fluke of nature may have allowed two incompatible species to breed, thereby giving birth to a unique creature such as that which has become known as the "Ringdocus".

For an explanation on anything, the 'Other' option should never be neglected; this one word covers every angle, both known and unknown. One possible suggestion that could fall into the other category is merely an undiscovered animal. Perhaps the 'Ringdocus' still exists today but has never been captured and analyzed.

The final, and possibly the most plausible explanation for the existence of the "Ringdocus" is that of a surviving prehistoric or Ice Age creature. A variety of primitive creatures have been considered a possible match for the 'Ringdocus' or 'Shunka Warak'in' [2,4,6,8] and their similarities and differences have been compared with the first suspected Shunka Warak'in sightings of 1886, as well as those sightings between 2005 to 2007 involving "The Creature of McCone County."

It is important to realize that the characteristics of the discussed primitive animals represent only those of that animal at the time of its common existence. If the creature did indeed survive to modern times, it would be logical to assume that they have also evolved (even though their evolutionary stages would not have been observed, and thus not appear within any scientific texts) and thus may present different characteristics than their primitive ancestors.

Similarities of the 'Ringdocus' with extinct animal species:

Bear Dog [13]

Fossil range: Eocene - Miocene
Scientific Classification
Kingdom: Animalia
Phylum: Chordata
Class: Mammalia
Order: Carnivora
Suborder: Caniformia
Family: Amphicyonidae
Subfamilies
Amphicyoninae   Daphoeninae   Thanmastocyoninae

Bear dogs are extinct Carnivoran mammals of the Amphicyonidae family. They lived during the Late Eocene to Late Miocene epochs. Physically, bear dogs were approximately as tall as that of the American black bear and were believed, due to the structure of their legs, to have ambushed their prey using short, sudden bursts of speed. Bear dogs have been incorrectly associated with the Waheela (a.k.a. Kerit)[3,14], a ghostly, snow-white wolf-like cryptid which has been reported within the Northwestern Territories of Canada. Although the Waheela is similar to the Shunka Warak'in, it inhabits a far more northern habitat and is not considered associated with the Shunka Warak'in in this work.

Borophaginae [15]

Fossil range: Oligocene to Pliocene epochs
Scientific Classification
Kingdom: Animalia
Phylum: Chordata
Class: Mammalia
Order: Carnivora
Family: Canidae
Subfamilies:

- Archaeocyon
- Oxetocyon
- Otarocyon
- Rhizocyon
- Tribe Phlaocyonini
  - Cynarctoides
  - Phlaocyon
- Tribe Borophagini
  - Cormocyon
  - Desmocyon
  - Metatomarctus
  - Euoplocyon
  - Psalidocyon
  - Microtomarctus
  - Protomarctus
  - Tephrocyon
- Subtribe Cynarctina
  - Paracynarctus
  - Cynarctus
- Subtribe Aelurodontina
  - Tomarctus
  - Aelurodon
- Subtribe Borophagina
  - Paratomarctus
  - Carpocyon
  - Protepicyon
  - Epicyon
  - Borophagus   (=Osteoborus)

The subfamily Borophaginae (hyena-like dogs) is an extinct group of canids which have hyena-like features and were endemic to North America. They have existed between the Oligocene to Pilocene epochs (40 to 2.5 million years ago).

It is believed that the Borophaginae descended from the Hespercyoninae subfamily. These animals have evolved to much larger sizes than their predecessors. Due to their powerful teeth and jaws, these predatory creatures have been generically referred to as "bone-crushing dogs." [15,16]

Although the description 'hyena-like dogs' closely resembles the basic physical characteristics of the 'Ringdocus,' a comparison of the teeth structures between the two creatures should be considered for a proper classification. Throughout my research, I have found no descriptions of the 'Ringdocus's' teeth. If they were of the 'bone-crushing' structure, they would be listed among the characteristics of the unknown creature. Due to this solitary fact, the 'Ringdocus' is unlikely to be part of the Borophaginae subfamily.

Cave Hyena [17]

Fossil range: Middle to Late Pleistocene
Scientific Classification
Kingdom: Animalia
Phylum: Chordata
Class: Mammalia
Order: Carnivora
Family: Hyaenidae
Genus: Crocuta
Species: C. crocuta
Subspecies: C. c. spelaea

The Cave Hyena (Crocuta crocuta spelaea) is an extinct subspecies of the spotted hyena (Crocuta crocuta) which is native to Eurasia. Cave Hyenas occupied the regions from Northern China to Spain and the British Isles. Although the Cave Hyena was originally considered a separate species from that of the modern spotted hyena, due primarily to large differences in the fore and hind extremities between the animals, genetic analysis reveals no significant differences in DNA between them.

Cave Hyenas were highly successful predators whose diet consisted primarily of Przewalski's Horses (16-51% of diet), the woolly rhinoceros (25-30% of diet), Reindeer (7-15% of diet), the Steppe Wisent (1 – 6% of diet), Red deer (3% of diet), rare instances of Irish elk, and chamois and ibex (representing less than 3% of diet). Cave Hyenas were also known to practice cannibalism.

Although, the Shunka Warak'in is reported to have hyena (or cave hyena) -like characteristics, the physical differences as well as the geographical regions between the two differ greatly. Thus, the possibility of the Shunka Warak'in being a descendant of the cave hyena is not considered within this work.

Chasmaporthetes [18]

Fossil range: Late Miocene to Late Pliocene
Scientific Classification
Kingdom: Animalia
Phylum: Chordata
Class: Mammalia
Order: Carnivora
Family: Hyaenidae
Genus: Chasmaporthetes
Species: Chasmaporthetes ossifragus

Chasmaporthetes are among one of several extinct hyena genera that hunted their prey at high speeds and possessed slender teeth and legs much like that of a modern African cheetah. These animals appeared approximately ten million years ago and became one of the few hyaenid genera to successfully compete with the canids in Europe and North America. The Chasmaporthetes reined over the regions that are now Arizona and Mexico from 5 to 1.5 million years ago. The species became extinct in the Pleistocene epoch during the Ice Age.

Since this genus primarily settled in Southwestern North America and Mexico, it is doubtful (but not impossible) to believe that neither the Hutchins's beast, the reported sightings of 1995, nor "The Creature of McCone County" would be surviving members of the Chasmaporthetes. However, the Hutchins's beast did present similar physical characteristics and may have represented a stray that ventured far from its common region or possibly represented a hybrid with a canid.

Dire Wolf [19]

Fossil range: Middle to Late Pleistocene
Scientific Classification
Kingdom:	Animalia
Phylum:	Chordata
Class:	Mammalia
Order:	Carnivora
Family:	Canidae
Genis:	Canis
Species:	C. dirus
Binomial name:	Canis dirus

The Dire Wolf (Canis Dirus) is an extinct carnivorous mammal of the Canis genus. This species was most common in North America during the Pleistocene epoch (1.6 million to 10,000 years ago). Although the Dire Wolf was closely related to the modern Gray Wolf, it was not the direct ancestor of any known present species.

The Dire Wolf and the Gray Wolf co-existed in North America for nearly 100,000 years.

The Dire Wolf was slightly larger than the Gray Wolf, averaging 5ft (1.5m) in length and weighing approximately 125-175 lbs. (57- 79 Kg). Although the Dire Wolf and Gray wolf were similar in appearance, there were significant differences between the species. The Dire Wolf had a smaller brain case than that of the Gray Wolf. Additionally, the legs of the Dire Wolf were proportionally shorter and sturdier than those of the Gray Wolf. It is believed that the stouter legs of the Dire Wolf adapted out of the necessity to develop power over speed to prey upon the slower-moving herbivorous animals that comprised the majority of their diet. Then, once these prey animals became extinct, the Dire Wolf may not have been able to chase down the faster prey that remained and thus had to rely upon carrion to survive.

There exists a link between that as the animals became slower moving, they also became more scavenger-like in nature, and thus developed larger, bone crushing teeth to acquire sustenance from the consumed carrion. Thus, the Dire Wolf also developed larger teeth than that of the more nimble Gray Wolf.

According to Inuit mythology, Amarok is the name of a gigantic wolf that would hunt down and devour anyone foolish enough to hunt alone at night. Amarok is rumored to hunt alone and is assumed to be a living Dire Wolf [20].

The stout legs of the Dire Wolf relate to the slower running speed of the animal; thus, it is unlikely for a researcher to associate Hutchins's beast with a surviving Dire Wolf, since the creature was rumored an extraordinarily fast traveler. It is also unlikely to consider "The Creature of McCone County" a Dire Wolf. As the Dire Wolf is similar in appearance to the Gray Wolf, it would prove more logical for the Montana Wildlife officials to misidentify the animal as a gray wolf rather than not being able to identify it at all. Therefore, neither Hutchins's beast nor "The Creature of McCone County" will be considered as a Dire Wolf.

Hyaenodon [21]

Fossil range: Late Eocene to Early Miocene
Scientific Classification
Kingdom: Animalia
Phylum: Chordata
Class: Mammalia
Order: Creodonta
Family: Hyaenodontidae
Genus: Hyaenodon
Species:

H. bavaricus	H. brevirostris	H. chunkhtensis
H. crucians	H. eminus	H. exicuus
H. gigas	H. horridus	H. incertus
H. leptorhynchu	H. megaloides	H. microdon
H. milloquensis	H. mongoliensis	H. montanus
H. mustelinus	H. pervagus	H.raineyi
H. venturae	H. vetus	H. weilini
H. yuanchensis		

The hyaenodon ("hyaena-toothed") [21,22] is an extinct genus of Hyaenodonts. These early carnivorous mammals were one of the latest genera of the Hyaenodonts. The Hyaenodon were known to exist from the late Eocene (55 to 38 million years ago) to the early Miocene (24 to 5 million years ago) epochs in regions of North America, Europe, Asia, and Africa. The North American Hyaenodon disappeared in the late Oligocene epoch (38 to 24 million years ago).

These animals had immense skulls, but small brains [21,23]. Their skull, much like that of the Shunka Warakin, was long and was characterized with a narrow snout. The body of the Hyaenodon was long and robust with a short neck, a low lumbar (back) region, and a long tail.

The average weight of the adult or subadult Hyaenodon horridus, the largest North American species, is estimated to be 88 lbs. (40 Kg) to 132 lbs. (60 Kg). It is possible that the "Ringdocus" may have been a surviving hyaenodon, as a close physical resemblance between the Hyaenodon and Hutchins's beast exists.

Long-nosed Peccary (Mylohyus nasutus) [24]

Fossil range: Pliocene – Early Holocene
Scientific Classification
Kingdom: Animalia
Phylum: Chordata
Class: Mammalia
Order: Artiodactyla
Family: Tayassuidae
Genus: Mylohyus
Species:

| M. elmorei | M. Floridness | M. fossilis |
| M. gidleyu | M. nasutus | M. sp |

Peccaries are members of the artiodactyls family Tayassuidae (Pigs and boars). A peccary can be easily distinguished from true pigs by observing the direction in which the tusks point. Peccaries' tusks' point downward, contrary to that of a true pig (whose tusks point upward). Additionally, peccaries are characterized by less complex cheek teeth, reduced side toes, and large dorsal musk glands.

The long-nosed peccaries were present in the Midwestern United States (as validated by fossil remains) up to only 11,500 years ago. Peccaries are believed to have first evolved around 33 million years ago in either North America or Eurasia. They then migrated into South America approximately 2.5 million years ago. Today, three species of peccary remain:

- Chacoan peccary (Vatagonus Wagneri) – found within the Gran Chaco regions of Bolivia, Paraguay, and Argentina.
- Collared peccary (Tayassu tajacu) – found along Arizona to Texas and into Northern Argentina.
- White lipped peccary (Tayassu peccary) - found from Southern Mexico to Northern Argentina.

A long-nosed peccary stood approximately 30 inches tall at the shoulder and is assumed to have weighed around 110 pounds. It is believed that the animals' diet was primarily comprised of shrubs, but probably included small invertebrates, nuts, and roots as well.

Peccaries are considered relatively primitive. It is assumed that the long-nose peccary was most likely a solitary animal and did not frequent caves, although their remains have been discovered within caves.

The long-nosed peccary most closely resembles the creature labeled "Ringdocus." First of all, the physical appearance of the recovered Ringdocus as seen in the photograph bears many similarities to the long-nose peccary. Furthermore, both creatures have incisors that point downward, stand approximately the same height, were considered solitary animals (i.e. only one "Ringdocus" was shot, and not a pack), were discovered within the same geographical area, and finally, were each considered primitive in nature (as witnessed by the savage feeding frenzy of a Shunka Warak'in) and the "Ringdocus".

In fact, the only apparent differences between the two animals are their diet and posterior features. However, a logical explanation can account for both. Because the remains of long-nose peccaries were discovered within caves, perhaps there were some peccaries that never ventured far outside of their habitat, for food was plentiful within the cave. These peccaries may have preyed exclusively upon cave dwelling animals or animals that had entered their lair and thus altered their diet accordingly. The posterior alterations could be attributed to evolutionary changes upon the animal or perhaps due to crossbreeding with different animals, as the majority of their own species would have become extinct.

Thylacine [25]

Fossil range: Early Pliocene to Holocene
Scientific Classification

Kingdom: Animalia
Phylum: Chordata
Class: Mammalia
Infraclass: Marsupialia
Order: Dasyuromorphia
Family: hylacinidae
Genus: Thylacinus
Species: T. Cynocephalus
Binomial name: Thylacinus Cynocephalus

The thylacine (Latin for Wolf-headed pouched dog) was the largest known carnivorous marsupial, which is believed to have only become extinct in the 20th century. This animal was commonly known as the Tasmanian Tiger, Tasmanian Wolf (a.k.a. Tassie or Tazzy Tiger) or simply the Tiger and was native to Australia and New Guinea.

The thylacine resembled large, shorthaired dogs with a stiff tail. Their coat was yellow-brown in color and featured 13 to 21 distinctive dark stripes running from its back to the base of their tail. Fur color varied from a light fawn to a dark brown, while the belly of the animal was cream colored.

A mature thylacine ranged from 39 inches (100 cm) to 71 inches (180 cm) long as measured from its' nose to the end of the tail and stood about 24 inches (60 cm) as measured from the front paw to the shoulder. The female thylacine had a pouch containing four teats which opened to the rear of the body, and the males had a scrotal pouch into which they could withdraw their scrotal sac.

The thylacine diet included kangaroos, wallabies, wombats, birds, and small animals such as potoroos and possums. It is believed that their favorite prey animal may have been the Tasmanian Emu, a large flightless bird which was hunted to extinction around 1850.

Thylacines were described as having a stiff and awkward gait, which made it unable to run at high speeds. However, the animal could also perform a bipedal hop, similar to that of a kangaroo, as an accelerated form of motion should the animal become alerted.

Although the Shunka Warak'in has been claimed by some people to appear as a surviving thylacine, there is little resemblance. Whereas the Shunka Warak'in has been commonly described as a wolf or dog hybrid with a hyena, the thylacine most clearly resembles a tiger. Furthermore, the native location, diet, and traveling speeds of the thylacine highly contradict the characteristics of the suspected Shunka Warak'in sightings. Thus, it is logical to dissociate the Shunka Warak'in with the thylacine.

Misidentified Shunka Warak'in kill sites

Before suspecting that a Shunka Warak'in is the culprit behind your missing/wounded/dead livestock, it is advisable to first consider the following modern predatory animals at fault:

Badgers [26]

A Badgers' diet includes rodents such as mice, prairie dogs, pocket gophers, young rabbits, ground squirrels, and occasionally includes small lambs and poultry. Badgers are known to den in crop fields. Although a badgers' tracks appear similar to those made by a coyote, close examination reveals that a badgers' tracks are distinctively "pigeon-toed" and may leave impressions on the ground from their long toenails.

Bears-Black & Grizzly [26]

Bears prey on livestock. The scene of a bear feeding is characterized by the torn, mauled, and mutilated carcasses of their prey. The victim is usually opened ventrally to consume the heart and liver. Smaller livestock animals such as sheep and goats may be consumed almost entirely, leaving only the rumen, skin, and large bones. If the prey is killed in the open, it may be moved to a more secluded spot for consumption.

Bear tracks resemble those of a human, but have distinctive claw marks. The print may appear four-toed if it was made in dust or shallow mud because the small inside toes do not often leave marks.

Bobcats and Lynx [26]

Bobcats and Lynx commonly prey upon smaller animals such as porcupines, poultry, rabbits, rodents, birds, and house cats; however, they occasionally prey on sheep, goats, deer, and pronghorns.

Coyotes, Wolves, and Dogs [26]

Coyotes, wolves and dogs prey on big game, livestock, rodents, wild birds, and poultry. Coyotes present the highest threat to livestock in the United States. These animals normally kill their prey with a bite to the throat, but have also been known to pull the animal down by attacking its' side, hindquarter, or udder. Coyotes have been known to remove the rumen and intestines and drag them away from the carcass of their prey.

Wolves prey upon larger animals such as caribou, moose, elk, and cattle. A Wolf generally brings down its prey by attacking the muscles and ligaments of the back legs. The fallen animals are then usually disemboweled.

Domestic dogs can present a serious problem to livestock, especially to sheep pastured near cities and suburbs. Dogs often attack the hindquarters, flanks, and heads of their prey. The attacking dogs often severely mutilate their victims but consume little flesh.

Dog tracks are round with their toes spread apart. Toenail marks can usually be seen on each of the toes. Dog tracks differ over those of the coyotes; whereas dog tracks are staggered, coyote tracks appear in a straight line.

Domestic Cats [26]

Domestic cats rarely prey upon animals larger than ducks, pheasants, rabbits, or quail. They tend to be untidy eaters, dispersing parts of their prey over several square yards. The large, meaty portions of their kill are consumed entirely, leaving only loose skin and scattered feathers behind.

Foxes-Gray and Red [26]

Gray and red foxes feed mainly on rabbits, hares, small rodents, poultry, birds, and insects as well as fruits. The gray fox also eats fish, which is seldom consumed by the red fox. Although both foxes are known to kill young livestock, poultry are their preferred diet. Foxes typically attack the throats of lambs and birds, but are also known to kill some by multiple bites to the neck and back. Normally, foxes take their fowl away from the kill site to their den.

Foxes consume the breast and legs of birds first and scatter their other appendages about which are then left partially buried.

Foxes dig their dens in wooded areas or open plains.

Hogs [26]

Wild hogs prey upon young sheep and goats. Typically, the entire carcass will be eaten at the scene or be carried off to be consumed at another location.

Mountain Lions [26]

Mountain Lions prey on deer, elk, and domestic stock, chiefly horses, sheep, goats, cattle, rodents, and other small mammals. A lone mountain lion is capable of slaying large numbers of animals in a single night. The mountain lion attacks its prey by inflicting powerful bites from above the victim, often breaking the vertebral column or neck. Lions usually prefer to first feed upon the front quarters and neck region of their prey. Their partially consumed meal is then carried off into the bushy areas and covered with litter to be hidden until the next feeding, which may last three to four nights.

Adult lion tracks are approximately 4 inches (10 cm) in length and 4 1/4 inches (11 cm) in width. Their prints have four well-defined toe impressions at the front of their paw and no claw marks.

Opossums [26]

Opossums are omnivorous animals eating fish, crustaceans, insects, mushrooms, fruits, vegetables, eggs, and carrion. They are known to raid poultry houses, killing only one chicken at a time. They consume young poultry and game birds entirely and only leave a few wet feathers behind.

Raccoons [26]

Raccoons eat mice, small birds, snakes, frogs, insects, crawfish, grass, berries, acorns, corn, melons, crops, etc. They have been known to kill small lambs by chewing their noses'. Raccoons may also enter a poultry house and attack several birds in one night.

The raccoon leaves a distinctive five-toed track that is similar to a small human handprint.

Skunks [26]

Skunks primarily consume insects, particularly grasshoppers, beetles, and crickets. When a skunk does attack poultry, it generally only kills one or two birds at a time and mauls them considerably. Skunk dens have been found to contain rabbit, chicken, and pheasant carcasses that have been dragged to the site as carrion.

Snakes [26]

Animals consumed by snakes may be quite hard to identify because they consume their prey whole, and generally leave no evidence behind. If you encounter an unaccountable disappearance of a small animal, a snake may be suspected.

Weasels and Mink [26]

Weasels and mink kill their prey by biting through the skull, upper neck, or jugular vein of their victim. They often kill many birds by eating away only their heads. When eating large muskrats, the mink eats away the flesh and pieces of the adjacent hide as the animal is skinned.

Summary

The Shunka Warak'in is described as resembling a cross between a wolf and a hyena. This creature is believed to occupy the wilds of the upper Midwestern United States. It is rumored to be a carnivorous predator that carries off dogs and viscously attacks livestock animals. The only known specimen of the Shunka Warak'in measures 48 inches long, 27-28 inches high, has a lupine head, narrow snout, sloping back, short hind legs, and is covered in fur consisting of light tan, brown, and black colors.

The first sighting of this mysterious creature occurred in 1886, when an unusual wolf-like animal was spotted within the Madison River Valley region of Montana. This creature was later shot dead by Israel Ammon Hutchins one morning in late January, and the carcass was donated to Joseph Sherwood to be displayed as a taxidermal exhibit in his combination grocery store and museum at Henry Lake in Idaho. Mr. Sherwood labeled this new exhibit "Ringdocus." Unfortunately, this exhibit became missing when it was sent to an unspecified location in the West Yellowstone area. However, the "Ringdocus" later resurfaced at the Idaho Museum of Natural History in Pocatello, in December of 2007. This animal has since been renamed "Guyasticuts" meaning 'Rocky Mountain Hyena,' and is currently undergoing DNA testing.

The second suspected Shunka Warak'in sightings occurred in 1995 within the North American states of Nebraska, Iowa, and Illinois, and in Alberta, Canada. However, little is actually known of this incident.

The latest suspected Shunka Warak'in activity occurred in late October of 2006 when an unusual wolf-like animal, later known as 'The creature of McCone County,' killed 120 livestock animals within the Montana counties of McCone, Garfield, and Dawson. This creature was later killed from the air by Montana's Wildlife Service agents on November 2, 2006.

Laboratory results from the creatures' DNA revealed a wolf-dog hybrid created as the result of a private captive breeding experiment. Furthermore, the laboratory was able to confirm that this hybrid creature was responsible for the killing of approximately 100 sheep in Garfield and McCone counties in 2006, thereby revealing the true identity of "The Creature of McCone County."

The only known physical specimen of the Shunka Warak'in ever examined was the taxidermal exhibit called "Ringdocus" or "Guyasticuts". It is believed that this physical specimen can be classified into one of the following categories:

- Merely a poorly mounted wolf
- A hand-stitched creation
- A unique genetic anomaly
- A truly successful natural hybrid
- A surviving prehistoric or Ice Age creature
- Other

When considering the possibility that the Shunka Warak'in is a surviving prehistoric or Ice Age creature, it is important to realize that the characteristics of the fossil remains of primitive animals only represent those of that animal at the time of its common existence. Any evolutionary changes that may have occurred within the creature to allow it to exist to present day would not have been recorded by evolutionary scientists. Thus, should a primitive creature still be in existence today, its characteristics would understandably be different from those which have been discovered through fossil remains.

The Shunka Warak'in has been hypothesized to be associated with or directly related to the following animals:
From mythology:

- Amarok,
- Waheela,

From the extinct animal species:

- Bear Dog,
- Borophaginae,
- Cave Hyena,
- Chasmaporthetes,
- Dire Wolf,
- Hyaenodon,
- Long-nosed Peccary (Mylohyus nasutus), and the
- Thylacine

Finally, it is well advised to first familiarize yourself with modern predatory animals and their feeding habits within your environment before claiming that a Shunka Warak'in is responsible for killing your livestock.

Conclusion

Of all of the known cryptids, the mystery behind the Shunka Warak'in is most likely to be soon uncovered. Unfortunately, for the Shunka Warak'in, it has been shot dead at every reported sighting. Probably, the prevailing justification for human hostility toward this creature is due to its appearance and behavior. Unlike the Bigfoot cryptid that has been reported as having a shy demeanor by fleeing from human encounters and appears humanoid in its silhouette, the Shunka Warak'in appears as a wild dog or wolf and is seen as a threat to livestock animals. Thus, although a hunter or farmer may express restraint before shooting a fleeing, non-threatening, animal or one that may appear humanoid in shape, there exists far less compassion for firing at a wild dog or wolf that is threatening their livestock animals.

Although little information has been uncovered about the Shunka Warak'in sightings of 1995, they are likely early hybrid experiments associated with the animal known as "The Creature of McCone County". In any account, the DNA of the "Ringdocus" shall reveal the true nature and origin of Hutchins's beast and those results should soon be published.

References

1. Cryptozoology A To Z: The Encyclopedia of Loch Monsters, Sasquatch, Chupacabras, and Other Authentic Mysteries of Nature, Authors : Loren Coleman & Jerome Clark, Copyright 1999, ISBN-13: 978-0684856025

2. Shunka Warakin - http://en.wikipedia.org/wiki/Shunka_Warakin

3. Cryptomundo.com - http://www.cryptomundo.com/cryptozoo-news/dire-waheela/

4. Nationmaster.com - http://www.nationmaster.com/encyclopedia/shunka-warakin

5. The Cryptid Zoo: American Hyena - http://www.newanimal.org/amhyena.htm

6. Unexplained-mysteries.com - http://www.unexplained-mysteries.com/forum/lofiversion/index.php/t47194.html

7. Unknown explorers.com - http://www.unknownexplorers.com/shunkawarakin.php

8. Cryptozoology.com- http://www.cryptozoology.com/glossary/glossary_topic.php?id=240

9. BOZEMAN DAILY CHRONICLE - http://www.bozemandailychronicle.com/articles/2007/11/15/news/000monster.txt

10. A Montana Wolf Mystery & the Fury it Breeds - http://www.newwest.net/index.php/main/article/an_e astern_montana_wolf_mystery_and_the_fury_it_bre eds/

11. Garfield County Creature Confirmed Hybrid - http://outdoornewsdaily.com/index.php/archives/502 7

12. Dead Mermaid Found in the Philippines - http://urbanlegends.about.com/library/bl_mermaid_p hilippines4.htm

13. Bear dog - http://en.wikipedia.org/wiki/Bear_dog

14. Waheela - http://en.wikipedia.org/wiki/Waheela

15. Borophaginae - http://en.wikipedia.org/wiki/Borophaginae

16. PHYLOGENETIC SYSTEMATICS OF THE BOROPHAGINAE (CARNIVORA: CANIDAE), Authors: Xiaoming Wang, Richard H. Tedford, Beryl E. Taylor, BULLETIN OF THE AMERICAN MUSEUM OF NATURAL HISTORY, Number 243, Issued November 17, 1999

17. Cave hyena - http://en.wikipedia.org/wiki/Cave_hyena

18. Chasmaporthetes - http://en.wikipedia.org/wiki/Chasmaporthetes

19. Dire wolf - http://en.wikipedia.org/wiki/Dire_wolf

20. Amarok- http://en.wikipedia.org/wiki/Amarok_(wolf)

21. Hyaenodon -
http://en.wikipedia.org/wiki/Hyaenodon

22. Fact File: Hyaenodon -
http://www.abc.net.au/beasts/factfiles/factfiles/hyaen
odon.htm

23. Hyaenodon horridus -
http://www.keltationsart.com/hyaenodon.htm

24. Longed-nosed Peccary (Mylohyus nasutus) -
http://www.isgs.uiuc.edu/education/ice-age-
res/peccary.shtml

25. Thylacine - http://en.wikipedia.org/wiki/Thylacine

26. Livestock and animal predation identification -
http://icwdm.org/Inspection/livestock.asp

27. Return to the Ice Age: The La Brea Exploration
Guide -
http://www.tarpits.org/education/guide/flora/wolf.ht
ml

28. Planetopia -
http://www.planetopia.cz/upload/image/hyeny/chas
maporthetes1.jpg

Final Words

This work was written to introduce the reader to the interesting world of cryptozoology. It was my intention to validate my claim that cryptozoology is as credible as any other field of theoretical study.

Through my research on the Shunka Warak'in, I created hypotheses about the unknown creature and then attempted to theoretically justify those hypotheses using analyzed data and drawing relationships to what would be considered similar animals. In effect, I have used the known to make judgments on what is unknown. This explains the basic principle behind any theoretical study. So on this account, it does not matter if you are hypothesizing on the nature of black holes, Bigfoot, or any other theoretical subject. All of these fields of study should be viewed as theoretical science and be respected as such.

Also available from the author:

Should I Take the Shot?
Practical and Ethical Considerations for Shooting a Bigfoot

This work offers an investigative overview on the feasibility of killing this legendary creature to provide proof that it exists and to examine the corpse in order to determine why it has become endangered. Both sides of the debate are analyzed and include a discussion of their positive and negative aspects.

ISBN: 1-4392-0954-5

INVISIBLE HUMANS
HISTORY, THEORY, AND APPLICATION

Throughout the ages, man has sought to harness the power of invisibility. An invisible person can enter a room undetected, be ignored by hostiles, gather information from remote locations, or just avoid an unwanted social interaction. The techniques described within this work cover the principles of invisibility based on physical, psychological, hypnotic, yogic, and occult practices. These highly guarded secrets, which have been passed down through the centuries,
are finally revealed.

ISBN: 1-4392-0953-7